DIG DEEP!
Bugs That Live Underground

Cicadas

Linda Buellis

PowerKiDS press™

New York

Published in 2017 by The Rosen Publishing Group, Inc.
29 East 21st Street, New York, NY 10010

Copyright © 2017 by The Rosen Publishing Group, Inc.

All rights reserved. No part of this book may be reproduced in any form without permission in writing from the publisher, except by a reviewer.

First Edition

Editor: Sarah Machajewski
Book Design: Mickey Harmon

Photo Credits: Cover (sky) Severe/Shutterstock.com; cover (background) ifong/Shutterstock.com; cover (cicada) Mark McElroy/Shutterstock.com; pp. 3–4, 6, 8, 10, 12, 14, 16, 18, 20, 22–24 (background) isaravut/Shutterstock.com; p. 5 baka_san/Flickr.com; p. 7 Tropper2000/Shutterstock.com; p. 9 (background) E. A. Given/Shutterstock.com; p. 9 (cicada) Mary Terriberry/Shutterstock.com; p. 9 (inset) the cicada project/Flickr.com; p. 10 Steve Heap/Shutterstock.com; p. 11 Auscape/Universal Images Group/Getty Images; p. 13 Ch'ien Lee/Minden Pictures/Getty Images; p. 15 akiyoko/Shutterstock.com; p. 17 Richard Ellis/Staff/Getty Images News/Getty Images; p. 19 (background) FotoRequest/Shutterstock.com; p. 19 (inset) Tom Franks/Shutterstock.com; p. 21 wxin/Shutterstock.com; p. 22 Steve Byland/Shutterstock.com.

Cataloging-in-Publication Data

Names: Buellis, Linda.
Title: Cicadas / Linda Buellis.
Description: New York : PowerKids Press, 2017. | Series: Dig deep! bugs that live underground | Includes index.
Identifiers: ISBN 9781499420500 (pbk.) | ISBN 9781499420524 (library bound) | ISBN 9781499420517 (6 pack)
Subjects: LCSH: Cicadas–Juvenile literature.
Classification: LCC QL527.C5 B85 2017| DDC 595.7'52–dc23

Manufactured in the United States of America

CPSIA Compliance Information: Batch #BS16PK: For Further Information contact Rosen Publishing, New York, New York at 1-800-237-9932

Contents

They're Here!....................4
Around the World...............6
What Do Cicadas Look Like?.....8
The Cicada Life Cycle..........10
Grown Up......................14
Cicada Songs..................16
Commonly Known Cicadas......18
Cicadas and People............20
Cicadas Are Cool!.............22
Glossary......................23
Index.........................24
Websites......................24

They're Here!

The next time you go outside, look down at the ground. What's living below your feet? Many kinds of **insects** live there, including cicadas. Cicadas spend most of their life underground, only coming above the surface occasionally. Some kinds of cicadas appear once a year, during summer. Some kinds stay underground for 17 years. That's a long time!

When cicadas finally show themselves, you know it. Hundreds of thousands of bugs can **swarm** a single area—and they're really noisy. We may not see cicadas much, but studying how and where they live tells us a lot about these **fascinating** creatures.

These cicadas were living underground just a short time ago. Why did they come above ground? Read on to find out!

Around the World

Scientists think there are more than 2,000 known cicada **species**. They live all over the world, usually in places that have warm summers. Many kinds of cicadas live in the eastern half of the United States. Many more species are found in Australia and its island of Tasmania.

Cicadas live in deserts, grasslands, and forests. Their **habitats** must have trees or large bushes because that's what they feed on when they're young. Also, the soil can't be too wet. When cicadas do come above ground, you may find them covering the trees and bushes in your neighborhood.

Dig Deeper!

Cicadas live on every continent except Antarctica.

Although cicadas spend much of their time underground, trees play an important part in their survival.

What Do Cicadas Look Like?

Cicadas come in many colors. They're commonly brown, black, green, or olive. Some are a mix of these colors. Cicadas are medium to large bugs. Their body is thick and commonly between 0.8 and 2 inches (2 and 5 cm) long.

Cicadas have two pairs of wings that fold over their back. They have big, round **compound eyes**, as well as three simple eyes. They have short antennae, or feelers, and mouthparts on their head. Male cicadas have special parts called tymbals on their **abdomen**. The tymbals **vibrate** very fast, which produces a buzzing sound. Female cicadas don't have them.

It's easy to tell cicadas apart from other bugs. Look for their big eyes, thick body, and clear wings.

The Cicada Life Cycle

A cicada begins life as an egg. Female cicadas lay their eggs in trees. They use a special body part to make a cut in a twig, branch, or limb where their eggs will stay safe. The eggs hatch on the plant. The newly hatched cicadas fall to the ground shortly afterward.

Young cicadas are called nymphs. At first, they look like small white ants. After the nymphs fall or crawl from the tree, they burrow, or dig, into the ground. Nymphs feed on plant roots, using their mouthparts to suck juices from the roots. It may be years before the young cicadas see the surface again!

Dig Deeper!

When female cicadas cut into a branch to lay their eggs, it can harm or kill the branch. Small, weak trees might even die.

Nymphs have a special pair of legs designed for burrowing into soil and for holding on to trees.

Cicadas are very active underground, even though we can't see what they're doing. Cicadas spend years tunneling and feeding on plant roots. They're also growing! They molt several times, which is when an insect sheds its **exoskeleton** to grow a new one.

Cicadas stay underground for 2 to 17 years, depending on the species. After that time, they **emerge**. At this point, they're still nymphs, but they're on their way to becoming adults. The cicadas climb onto the nearest tree. There, they go through one final molt. The cicadas shed their exoskeleton. They're now adult cicadas. The exoskeleton they leave behind is proof they were there!

Dig Deeper!

The cicada's exoskeleton looks like a perfect outline of the cicada nymph.

Cicadas leave their old exoskeleton behind as they grow.

Grown Up

Adult cicadas look different from nymphs. The biggest difference is that they now have wings! An adult cicada's wings and exoskeleton harden after its last molt. Now, cicadas are ready to fly. They can crawl, too, but they can't jump like other insects do.

Adult cicadas live for about four weeks. They stick close to trees, looking for a mate. A mate is one of two animals that come together to make babies. After a male and female join together, the female lays eggs. The adult cicadas die shortly after. Soon, new cicadas will hatch, and the life cycle starts again.

Cicadas don't go far from where they molted into adults. Some cicada species stay together in large groups to mate, which is one reason why you can see so many of them in a single place.

Dig Deeper!

Adult cicadas use their mouthparts to suck juices from plant stems.

Cicada Songs

Male cicadas are the only cicadas that can "sing." That's because only males have the tymbals that vibrate to make noise. Male cicadas sing to attract females. They may also use their call to tell other cicadas that danger is nearby. Males may make a loud buzzing noise to surprise predators that get too close.

Each cicada species has its own call. To people, though, they may all sound the same. We hear cicadas' calls as buzzing or clicking. They're pretty loud! When hundreds of cicadas gather in the same area, all the singing can sound like a loud, buzzing hum.

Dig Deeper!

Female cicadas can make a noise by flicking their wings, but it's not the same as the males' songs.

Most cicadas sing during the day and into the early evening. Their songs can be much louder than other noises around them.

Commonly Known Cicadas

There are thousands of cicada species in the world. A type called the dog-day cicada is well known. These cicadas are annual. That means some of them come above ground once a year, during July and August. Some people call these months the "dog days" of summer, which is where this type of cicada gets its name.

Some of the most fascinating cicadas are the 13-year cicada and the 17-year cicada. These species belong to a group called *Magicicada*. They're named after the number of years it takes to complete their life cycle. One of the biggest broods, or groups, of 17-year cicadas came above ground in 2004. We're sure to see the next generation soon!

dog-day cicada

17-year cicada

It's amazing that some cicada groups appear once every 17 years!

Cicadas and People

Cicadas appear in such large numbers that they could be considered pests. However, they don't do much harm when they appear. They don't bite or sting people, but they are pretty loud. Cicadas can harm young trees when they lay their eggs, but older trees usually survive.

Cicadas have fascinated people for centuries. In ancient China, cicadas were seen as symbols of rebirth, or being born again. They were also captured for their song. Some people thought cicadas' songs signaled a change in the weather. Cicadas were also eaten by people of many cultures, including American Indians. In fact, people still eat cicadas today!

Are you brave enough to eat a cicada?

Cicadas Are Cool!

Living underground is the best way for cicadas to survive. Predators can't reach nymphs underground. Adult cicadas have many predators that eat them when they emerge, including birds, snakes, moles, lizards, and even fish! However, there are usually so many cicadas that predators can't catch them all. For this reason, cicadas are not **endangered**.

Cicadas will likely be around for many years to come, but we won't see them that often. Before they come out again, take a look at the ground. What are the cicadas doing under there?

Glossary

abdomen: The part of an insect's body that contains the stomach.

compound eye: An eye that has many parts.

emerge: To move out of something and come into view.

endangered: In danger of dying out.

exoskeleton: The hard outer covering of an insect's body.

fascinating: Very interesting.

habitat: The natural place where an animal, insect, or plant lives.

insect: A bug with six legs and one or two pairs of wings.

species: A group of living organisms that share similar traits and that can come together to make babies.

swarm: To move somewhere in large numbers.

vibrate: To move back and forth very quickly.

Index

A
American Indians, 20
Antarctica, 6
Australia, 6

C
China, 20

E
eggs, 10, 14, 20
exoskeleton, 12, 13
eyes, 8, 9

F
feelers, 8, 9

H
habitat, 6

M
Magicicada, 18
molt, 12, 14
mouthparts, 8, 10, 15

N
nymphs, 10, 11, 12, 14, 22

P
predators, 16, 22

R
roots, 10, 12

S
songs, 16, 17, 20
species, 6, 12, 14, 16, 18

T
trees, 6, 7, 10, 11, 12, 14, 20
tymbals, 8, 9, 16

U
underground, 4, 7, 12, 22
United States, 6

W
wings, 8, 9, 14, 16

Websites

Due to the changing nature of Internet links, PowerKids Press has developed an online list of websites related to the subject of this book. This site is updated regularly. Please use this link to access the list: www.powerkidslinks.com/digd/cica